EXTREME CHEATS

BLACKBIRCH PRESS

An imprint of Thomson Gale, a part of The Thomson Corporation

THOMSON

GALE

Detroit • New York • San Francisco • San Diego • New Haven, Conn. • Waterville, Maine • London • Munich

THOMSON
GALE

For more information, contact
Blackbirch Press
27500 Drake Rd.
Farmington Hills, MI 48331-3535
Or you can visit our Internet site at www.gale.com

Photo credits: Cover: all Corel Corporation except top center © Digital Stock; middle left © Photos.com; interior: all pages © Discovery Communications, Inc., except for page 1, 12, 26 © PhotoDisc; page 19 © Photos.com; pages 4, 8,16, 20, 24 Corel Corporation; page 28 © Tom Brakefield/CORBIS; page 40 © Eric and David Hosking/CORBIS; page 32 © Photos.com; page 36 © Digital Stock; page 38 (bottom) Library of Congress; page 39 © Royalty-Free/CORBIS; page 44 National Archives

LIBRARY OF CONGRESS CATALOGING-IN-PUBLICATION DATA

Cheats / John Woodward, book editor.
p. cm. — (Planet's most extreme)
Includes bibliographical references and index.
ISBN 1-4103-0381-0 (hardcover : alk. paper) — ISBN 1-4103-0339-X (paper cover : alk. paper)
1. Animal behavior—Juvenile literature. 2. Camouflage (Biology)—Juvenile literature. I. Woodward, John, 1958–. II. Title III. Series.

QL751.5.C46 2005
591.47—dc22

2004020637

Printed in the United States of America
10 9 8 7 6 5 4 3 2 1

Animal Planet THE MOST EXTREME

If you think you're sneaky, just wait until you see the dirtiest rotten scoundrels in the natural world. We're counting down the top ten most extreme cheats in the animal kingdom and seeing how they compare to human cons, crooks, and swindlers. For some animals, crime really does pay when it's taken to The Most Extreme.

10 The Fox

Sneaking in at number ten in the countdown is the fox. In the mountains of South America, the fox has to be sneaky because around here, the birds are big. The rhea is a flightless bird that's more than a match for even the hungriest fox, so the fox can't use a frontal attack. By waiting for the rhea family to move on, it is free to make an omelette out of the eggs that failed to hatch.

The only trouble is that big birds lay big eggs. But the fox is number ten in the countdown because it has a reputation for coming up with sneaky solutions to difficult problems. If you can't bite it, kick it. A sharp blow against a stone is enough to break the shell.

With the cunning fox around, it's no wonder that most birds keep their eggs in nests far from the ground. Not even the wiliest fox could reach eggs up here, in a tree. Or could it?

A fox smashes a large rhea bird egg to slurp up the tasty inside, yolk and all. No eggs are safe from a fox willing to climb the tallest tree!

With brains, agility, and good looks, the fox is the perfect superhero! "Zorro" is the Spanish word for fox, and the name of the swashbuckling hero of countless movies. The sneakiest thing about that human fox was his ability to hide his true identity using only a simple disguise.

Many movies feature the hero Zorro. "Zorro" is the Spanish word for fox.

El Zorro has a reputation as a very sneaky thief. You would, too, if your job was to steal eggs when no one was looking.

Few animals have a sneakier reputation than El Zorro. But even the wily fox can't escape the fact that it's only number ten in the countdown.

9

The Rat

If you live in garbage, you've got to expect people to talk trash about you. That's why the rat creeps in at number nine in our countdown.

We hate rats because they're such extreme cheats. They cheat us out of food, and they cheat us out of money. Each year, rats cause an estimated $19 billion worth of damage in the United States alone! Rats steal from right under our noses. It's estimated there are 70 million rats in New York. That's almost ten rats for every person in the city!

It's the rat's speed and agility that makes this cheat hard to beat. Rats will go to extreme lengths for a free meal. That's why some people love the clever little rodents. Debbie Ducommun is founder of California's Rat Fan Club. She explains why she likes them so much:

> *Rats have a bad reputation, but wild rats are just like any other wild animal. They're just trying to survive. It's not like they're trying to hurt us. It's just that they are smart enough and brave enough to figure out that they can live off of food we leave laying around.*

Ducommun is determined to clean up the rat's rotten reputation.

> *We go to pet fairs where we educate people about rats, and tell people how they really are instead of how they're shown on television and movies. Rats are very social animals, which means that they form strong social bonds. They really bond to people. They really know who you are and have a relationship with you. That's the main reason why I like them so much.*

Some gamblers try to cheat. This roulette player is secretly moving his chips after the winning number has been called.

The much-maligned rat may actually be closer to humans than we like to think. Deception comes naturally to all living things, including humans. According to one psychologist, chances are that you are lied to about 200 times a day! That's roughly one untruth every five minutes!

Why do we cheat? Nature rewards successful cheats by allowing them to survive long enough to reproduce. Just like the rest of the animal kingdom, humans cheat either to avoid trouble or to get something they can't get by other means. For some people, the thing they want most is money. More than 150 million people visit U.S. casinos each year, and some resort to cheating to beat the odds. A common con is to try to place a bet on a winner after a game is over. The cheat waits until the dealer is distracted, and then makes his move.

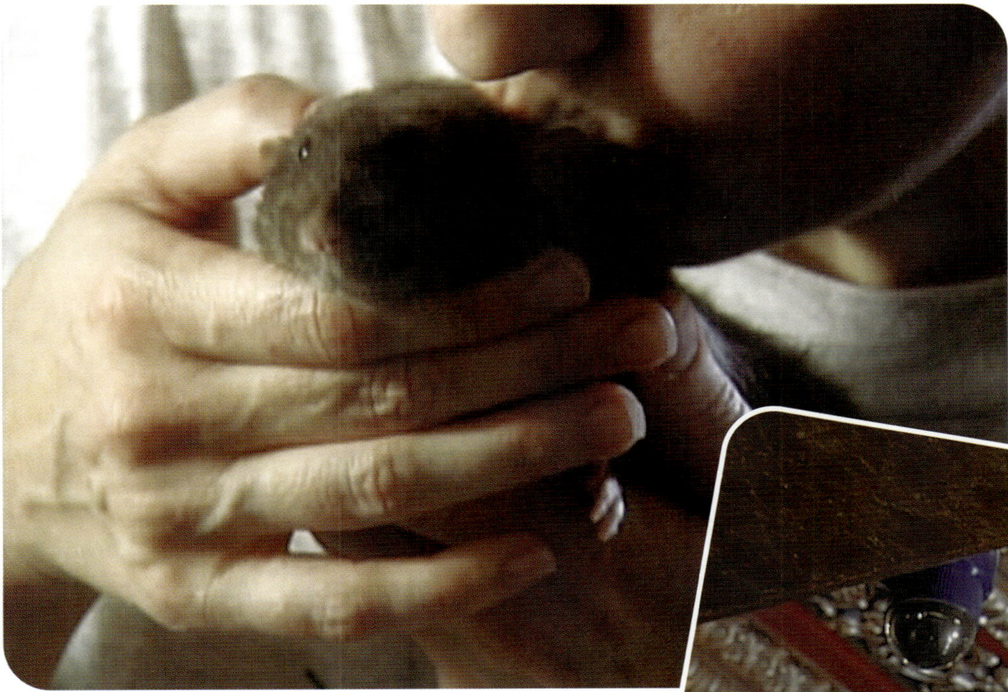

Casinos use security cameras to watch for cheats (inset). Although rats cheat more than any gambler, some people actually love them.

But the biggest casinos have literally thousands of security cameras waiting to catch the cheat in the act. Some cheats are caught trying to place a bet on the winning number after the result was known. They think they can get away with it by blocking the dealer's view of their cheating hand. But they usually can't beat the system. Perhaps if we took a closer look at rats, they could beat the odds and win their way into our affections.

8

The Chameleon

In parts of Africa, people are afraid of number eight in the countdown. They just don't trust the chameleon. In some languages the word for "chameleon" and "liar" is the same thing. As with all good liars, you can hardly believe what comes out of its mouth. The chameleon's tongue is like an elastic harpoon tipped with sticky chewing gum. How would you like a tongue that's twice as long as your body?

Most people distrust chameleons because they're masters of disguise. In a Los Angeles classroom, biological educator Dave Riherd conducts experiments to see if the chameleon really can change color to match even the most extreme background.

When its mood is good, the chameleon is camouflaged green. But if two males get together they really see red! It's easy to spot an angry chameleon—it's written all over its body. That's because its skin has special expandable cells full of color pigments. It's a shame human cheats don't send out such obvious warning signals.

A chameleon shows off its sticky tongue (top). While green is the color of a happy chameleon (middle), an angry one turns bright red (bottom).

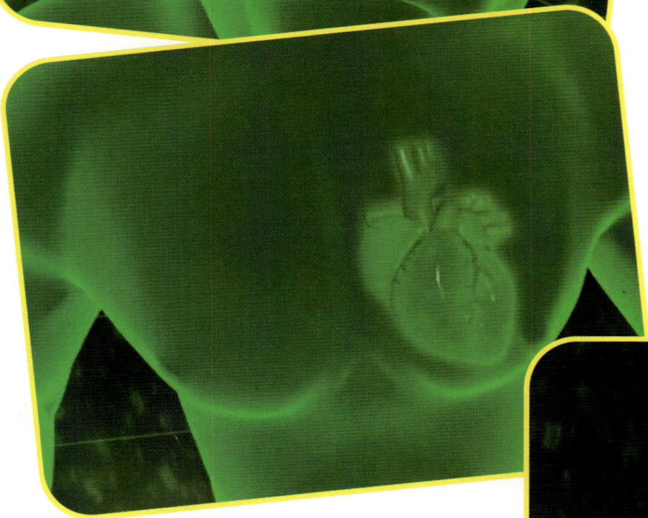

A good liar will never turn red. But there are subtle ways of seeing past even the best poker face. If you're looking for a human cheat, watch out for symptoms of anxiety like an increased heart rate, sweating, fidgeting, or nervous touches to the face.

You can spot even the best liars if you know what to look for. As they sweat like crazy, their heart beats faster, and they keep touching their face.

Scientific research shows the greatest giveaway may be in the voice. Anxiety causes muscle tension and reduces blood flow to the vocal cords. This produces a distinctive pattern of sound that can give away even the best cheats. Unfortunately, it often takes sophisticated voice-analysis equipment to detect good liars. Imagine if we were like chameleons and could tell what people were really thinking by their skin color!

The voice of a liar has a very distinctive sound. Just listen to old tapes of former president Richard Nixon speaking.

7 The Caterpillar

Number seven in the countdown, the caterpillar, is hard to beat, and even harder to eat. This cheat is a master of disguise. Even entomologist Ruud Kleinpaste has to work hard to see through the incredible camouflage of the caterpillar. He explains:

> ***It's not easy being a caterpillar. You're soft, you're juicy, and you're a nice one-bite meal. So they're masters of disguise. They make themselves look like anything but food.***

If you're on the bottom of the food chain, it pays to look like something that nobody wants to eat. Caterpillars are number seven in the countdown because of the vast array of tricks they use to fool hungry eyes. The bird-dropping caterpillar avoids being eaten by disguising itself as something that a bird has already eaten!

Other caterpillars are thieves. They steal poisons from the plants they eat and tip their spines with it. Some caterpillars pretend to be much tougher than they really are. A bird may not be so hungry if it thinks it's attacking a tiny snake!

The most extreme caterpillar on the planet lives in Hawaii. These are the only caterpillars in the world that eat flesh. Their cunning camouflage lets the hunted become the hunter.

A caterpillar disguises itself as bird droppings (top). Another tries to look like a tiny snake (middle), while a third carnivorous caterpillar (bottom) hunts for unsuspecting prey.

It's amazing what you can do with a good disguise. This stealth aircraft is the caterpillar of the air. It's so well disguised that it has the radar signature of a small bird rather than an airplane. That's how it can launch such devastating attacks.

Stealth technology means that the airplane's shape and skin materials are designed to minimize radar reflections. But this state-of-the-art technology is nothing new for one adult caterpillar. The tiger

Designed to fool radars, the stealth aircraft appears on radar as a small bird instead of the large plane that it is.

The tiger moth flies like a stealth bomber, using its fuzzy scales to absorb sonar signals given off by hungry bats.

moth has its own stealth technology on its wings. A covering of fuzzy scales absorb not radar, but sonar signals used by hunting bats. Just like a stealth bomber, the tiger moth is able to fly through the night sky invisible to the hungry bat's sonar signals. Thanks to cunning camouflage, moths and caterpillars may be able to stay off the menu.

6 The Turtle

Our next contender disguises its tongue not as a caterpillar, but as a worm. Like every fisherman, the snapping turtle is a cheat. He wants to fool a fish into taking the bait. Some cheats are better than others, however, which is why this turtle is number six in the countdown.

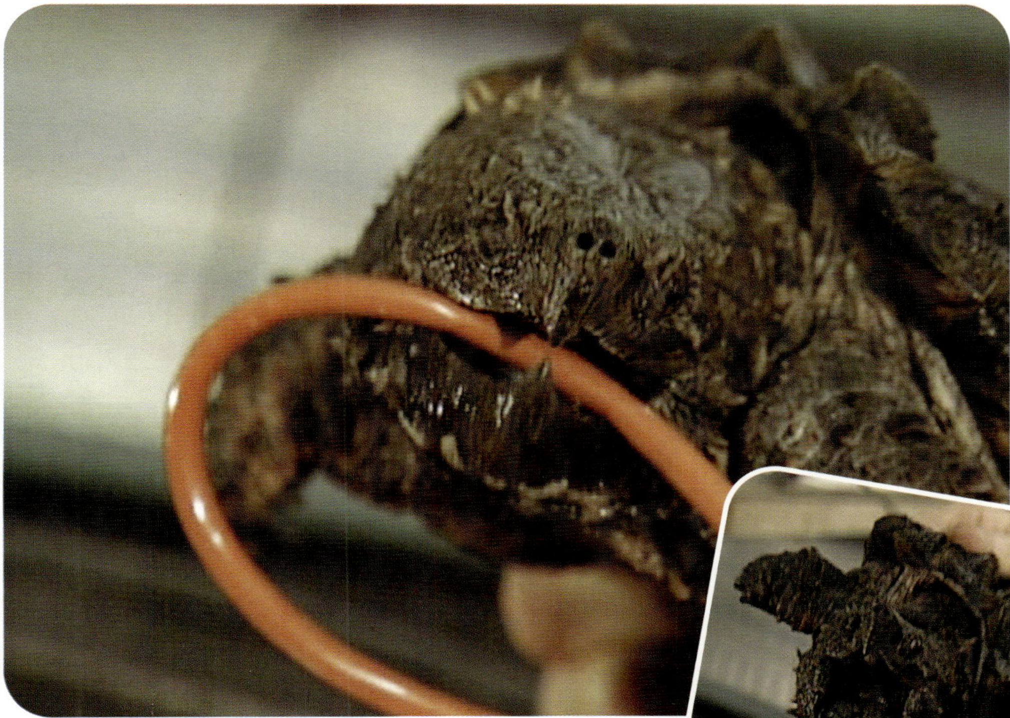

With its powerful jaws (inset), a snapping turtle shows off its super fast reflexes by biting a rubber tube as soon as it's placed in its mouth.

St. Louis zookeeper Peter Taylor explains:

This is a five-year-old alligator snapping turtle. He's a bit threatened about being out of the water, so he's gaping. See the wonderful profile of those raptor-like jaws? As a demonstration of the fast reflexes of this turtle, I'm going to take a soft object, this rubber tube, and see what happens when I touch the inside of the jaws. See how quickly he snapped it? An adult turtle could remove a finger.

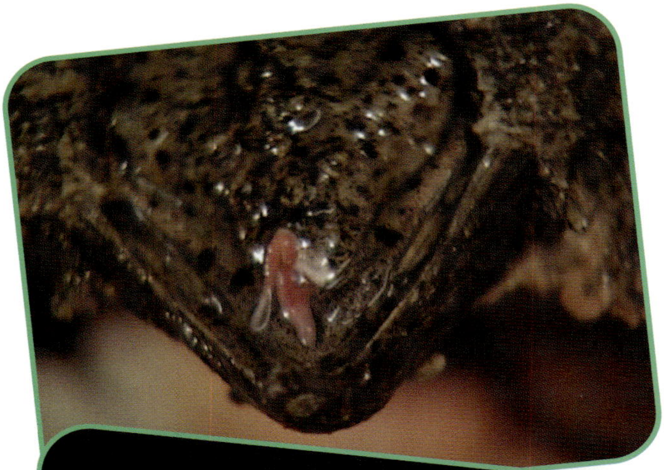

The tricky part is getting a fish to swim between these lethal jaws, and that's where this little pink worm comes in handy. Fish are easily fooled into believing that the turtle is just another log in the pond. To increase the chances of a fish swimming into the trap, the tip of the snapper's tongue does a remarkable impersonation of a tasty worm.

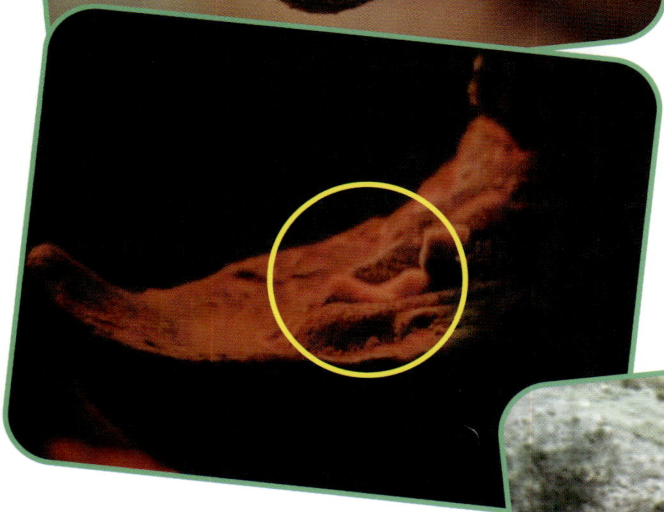

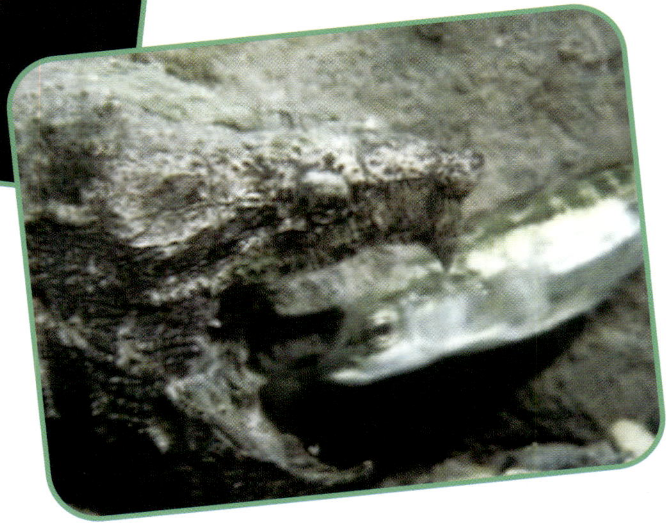

By imitating a worm with the tip of its tongue, a snapper lures in and snags a tasty fish with lightning speed.

Crooks in Texas used this snapping turtle as a weapon to rob a pizza delivery driver.

Some human criminals went one step further and used not only the worm, but the whole snapping turtle to go fishing for money! In Balch Springs, Texas, a pizza delivery man was robbed of 50 dollars. Two crooks cornered him in a phone booth and threatened him with an alligator snapping turtle!

5

The Firefly

The number five creature in the countdown, the firefly, knows that it pays to advertise. Jane Stevens of the St. Louis Zoo explains:

> *The female sits quietly and blinks from a leaf or a branch. It's like saying, "Hey, I'm over here." And then the males say "Hi, how are you?" and come over to visit. It's a wonderful way to communicate, and not one that's very common in the insect world.*

There are about 136 species of fireflies within North America and all of the species have their own individual glow. The shape of the glow is also important. Sometimes they actually acquire a shape like writing your name in the air with a sparkler at night on the Fourth of July.

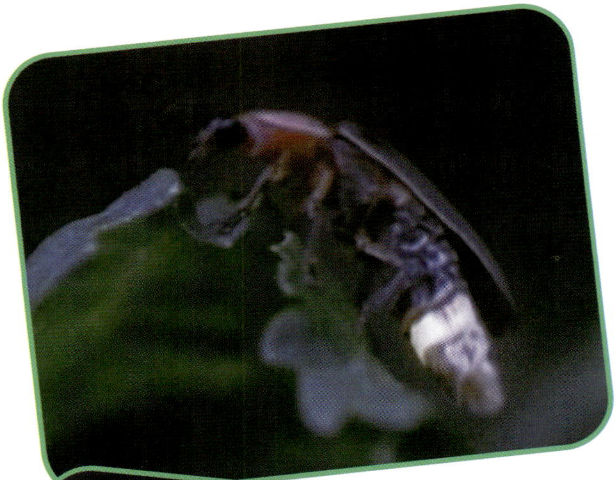

For anybody who can crack the code of the firefly, catching these bugs suddenly gets really easy. All you have to do is mimic their flashing signals and males will fly right into your hands.

Unfortunately for the males, there's one deadly female that uses a little trickery of her own. This female firefly has discovered how to mimic the flashing code of other species. She's able to send out an invitation to any males that might be passing by. The firefly femme fatale fools the male into thinking he's on a date, when he's actually just a hot meal!

A female firefly flashes her light to lure a male. Instead of mating with him, however, she has him for dinner!

Women often wear makeup to attract the opposite sex. Fortunately for men, most women don't eat their dates.

Fireflies are not the only animals that use visual signals to attract a mate. Putting on makeup is one way that human females highlight their best features to enhance their attractiveness to the opposite sex. Fortunately, few women will kill the males they attract.

For centuries people have been using artificial chemicals to help them win in the dating game. Perfumes not only smell nice, but they're thought to contain molecules that mimic pheromones, the odorless chemicals that make us attractive to the opposite sex. Some scientists believe that it's not love at first sight, but first smell.

Catching a whiff of someone's pheromones may be the reason why we find them attractive. And now it's possible to cheat a little. Researchers in Britain have come up with a preparation of human pheromones that you can use to increase your appeal. The manufacturer claims it will boost your pheromones and make you more popular on the dance floor!

Some people use perfume to smell nice for the opposite sex. Others prefer odorless pheromones to boost their appeal.

4

The Hanuman Monkey

Number four in the countdown once lived in the jungles of Southeast Asia. Then humans came along, and good monkeys went bad. Hanuman monkeys live a charmed life because their namesake is worshipped by millions of people. Hanuman is a powerful god and wouldn't take kindly to anybody hurting the animals that carry his name.

Unfortunately, some monkeys have taken advantage of their divine protection and get up to all kinds of monkey business. These privileged primates are number four in the countdown because they've become terrible thieves.

Tourists love watching the monkey's antics, until they discover they've lost not only their water and their lunch, but also handbags, sunglasses, and cameras. Locals are terrorized by street gangs of hungry monkeys, yet their religious beliefs require that they don't harm even the worst offenders.

Some monkeys have even learned to get food by extortion. They'll wait for people to hang their laundry out to dry. Then they pounce. If you try to scare the monkey to get your clothes back, it will run away and your clothes are gone forever. It's only when you throw the monkey food that you get your laundry back.

A Hanuman monkey steals a bottle of water from a tourist, while another runs off with someone's laundry.

Using threats to get what you want is nothing new. Some people made a career of it. Butch Cassidy and the Sundance Kid were perhaps the most legendary thieves in the world. They specialized in holding up trains, although sometimes things didn't quite go as planned. In 1899 they stopped a train in Wyoming. They boarded the train carrying more than enough dynamite to blow up the safe in the express carriage. The blast destroyed the carriage, but the safe was unharmed, so they used more dynamite. This time it worked so well that 30,000 dollars were blown into the sky!

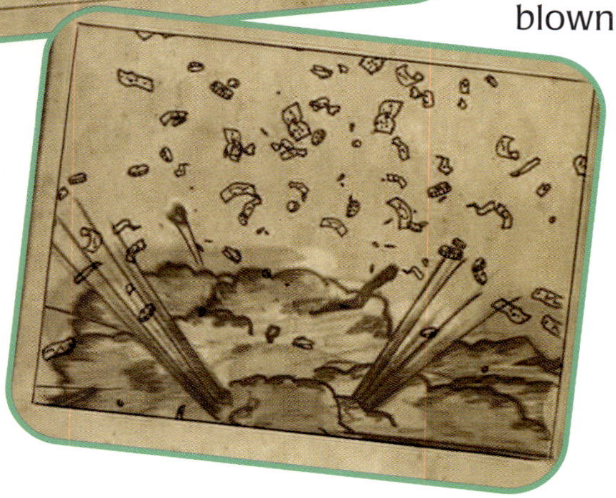

Butch Cassidy and the Sundance Kid were legendary train robbers who didn't always have bright ideas. They once used too much dynamite on a safe and blew up all the money!

Hanuman monkeys love thievery and mischief. They can even be trained to pick the pockets of hapless tourists.

It's fortunate that monkeys don't have access to dynamite. But some humans are training monkeys in the fine art of stealing. They've taken the monkey's natural talent for mischief and taught them to steal for their trainers! Forget Butch and Sundance—these monkeys are top banana when it comes to thievery.

3 The Opossum

Dogs can be taught to do just about anything. They can even learn to play dead—a trick that was inspired by the animal coming in at number three in the countdown, because to play dead is to "play opossum." Number three on our list is the opossum.

Nobody would ever accuse the Virginia opossum of being reckless. Rather than getting into fights, the opossum prefers to play dead.

The Virginia opossum is America's only native marsupial. It's not much bigger than a house cat, and it's a slow mover. Opossums will generally try to keep out of everybody's way. If they're cornered, however, they will play possum, which is a way of fooling even the hungriest predator. This deception has let them survive more or less unchanged for 70 million years.

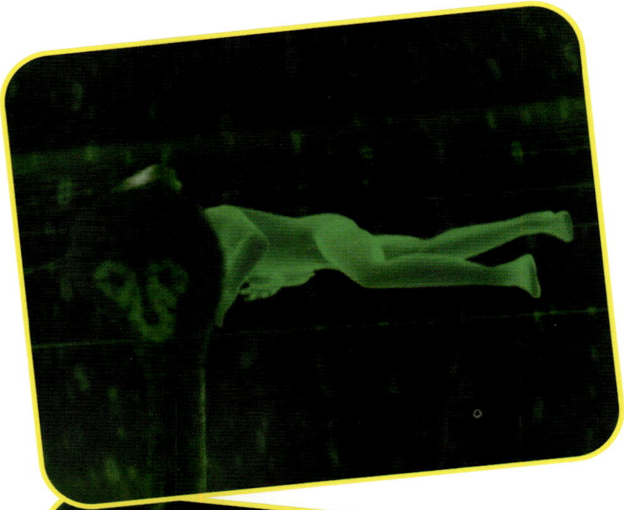

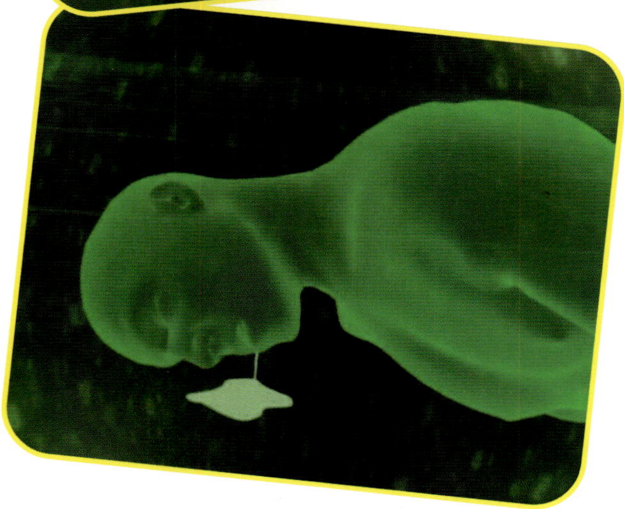

If you were like a scared opossum, you'd enter a catatonic state. That means every time you felt frightened, you'd drool and go into a coma.

If you could play dead like an opossum, you'd have no trouble winning the Academy Award for best performance. That's because opossums don't really "play" dead at all. If you were an opossum and you were in a dangerous situation, first you'd drool a little saliva and excrete a greenish, foul-smelling mucus, then you'd go into a voluntary coma! The opossum is not just pretending to be dead, it really goes into a catatonic state. If you were an opossum it would look as if you'd been frightened into cardiac arrest. You'd feel no pain no matter how much you were poked or prodded. Most predators like fresh meat. If they think you're already dead, they may also think you're rotten and full of nasty diseases, and will leave you alone.

Traveling medicine men like this one made money by selling cures for every ailment. People truly believe most anything you tell them!

Humans can cheat each other because we'll believe anything. This is why traveling medicine men were so successful. They would claim to have invented a miracle cure for any ailment even though they had no medical training. One such inventor made the most extreme medicine in the world. In 1886, John Penderton was working on a cure for headaches. His mixture didn't cure many headaches, but when he added it to soda water, he had created the recipe for Coca-Cola.

2 The Orangutan

The orangutan gets its name from two Malay words that mean "old man of the forest." Orangs are primates, and they can use tools to solve problems like we do. Because of this intelligence, orangs are number two in our countdown of extreme deceivers.

Spend any time with an orang and you'll quickly discover just how clever it is. Michelle Farmerie of the Pittsburgh Zoo explains:

> *Orangs are excellent problem solvers. One of our female orangs got a piece of rock and she used it to break free the drain cover at the bottom of the exhibit. It's a heavy piece of metal. She used that drain cover to try and knock the window bolts loose. We have to be very mindful of what we give our orangs because of their intelligence and their ability to manipulate their environment.*

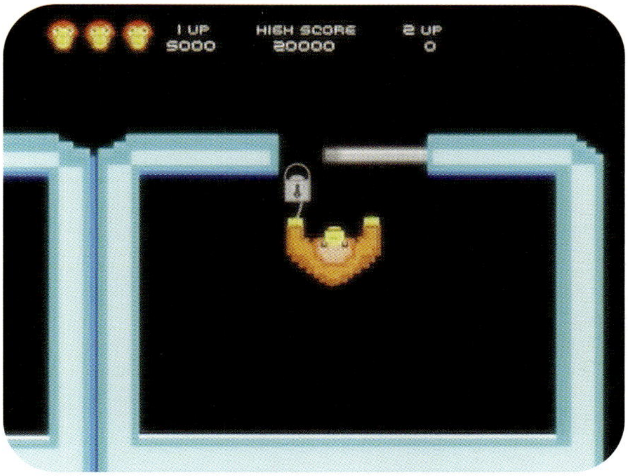

A video game of the lock-picking adventures of Fu Manchu the orangutan might look like this.

Many zookeepers use orangs to test new enclosures. If an orang can't get out, nothing can. A few years ago at the Omaha Zoo, one orang named Fu Manchu became so famous for opening locked doors that he became an honorary member of the Greater Philadelphia Locksmith's Association! Escaping from his enclosure became an exciting game Fu Manchu regularly played with his keepers.

Fu Manchu had secretly made himself a wire lock-pick and he kept it hidden by storing it between his lip and gum. He'd use it to open his door, and then he'd let all his friends out. While they ran about the zoo causing trouble, Fu would head back to his room and sit innocently waiting for the keepers to find him!

Not even the sneaky Fu Manchu could match the most extreme human escape artist of all time. Nothing could contain the incredible Harry Houdini. Houdini claimed he could escape from anything. He even offered rewards to anyone who could successfully restrain him. He escaped from such things as a burglar-proof safe, a piano, a coffin, and even a preserved giant squid!

Harry Houdini claimed he could escape from anything, and he accepted many challenges. He even escaped from the body of a giant squid!

The orangutan is the hairy Houdini of the animal world, always looking for an escape route.

We'll probably never find out for sure if an orang can escape from a preserved giant squid. Even these masters of deception, however, are still not the greatest tricksters in the countdown.

1 The Cuckoo

Imagine if your family was duped by the most extreme cheat on the planet. How would you like it if your baby were swapped for an infant that grows much, much bigger? If you were conned by the number one cheat in the countdown, you'd be left raising a baby that grows ten times bigger than you are! That's exactly what happens if you're conned by the number one cheat, the cuckoo.

An unsuspecting warbler checks on her eggs. Shortly after, a cuckoo eats one of the warbler eggs to make room in the nest for her own egg.

The scam begins when a female cuckoo raids the nest of a poor little reed warbler. First she has to get rid of a warbler egg. Then she replaces the warbler egg with one of her own.

The warbler thinks the cuckoo egg is one of her own, and she sits on it with the others and keeps it warm.

One reason the cuckoo is number one in the countdown is that it's a master of disguise. Its egg is so well disguised that it fools the reed warbler into believing it's one of her own.

The con doesn't end there—for baby cuckoos are nasty. As soon as it hatches, the cuckoo chick will get rid of all competition, and it does it right under its foster mom's nose! Then it literally eats its parents out of house and home. The poor reed warblers end up feeding a baby that's ten times heavier than they are! The warblers have no choice, because this baby is part of the mob.

A baby cuckoo pushes a warbler egg out of the nest (top). As a dutiful parent, the warbler feeds the enormous cuckoo chick (bottom).

Mobsters were big in Chicago in the 1920s. Al Capone's business card may have read "Second-hand Furniture Dealer," but in reality he was a gangster. He ruled Chicago.

This mobster attack on Mike's Barber Shop in Chicago was probably ordered by Al Capone, a gangster of the 1920s.

A cuckoo chick sits in a warbler nest. As The Most Extreme cheat, the cuckoo tricks other birds into caring for its babies.

Scientists discovered that cuckoos are the Capones of the bird world when researchers removed cuckoo eggs from magpie nests. A few days later, they returned to find that more than half the nests had been completely destroyed by the cuckoo mafia.

Cuckoos make birds an offer they can't refuse: Take care of my baby or suffer the consequences. No wonder then, that when it comes to treachery—the cuckoo really is The Most Extreme.

For More Information

Stephen Brend, *Orangutan*. Austin, TX: Raintree Steck-Vaughn, 2000.

Sneed B. Collard III, *A Firefly Biologist at Work*. New York: Scholastic, 2001.

Richard Conniff, *Rats! The Good, the Bad, and the Ugly*. New York: Crown, 2002.

Sandra Donovan, *A Fox in Its Den*. Minneapolis, MN: Lake Street, 2003.

Marjorie Facklam, *What Does the Crow Know? Mysteries of Animal Intelligence*. Layton, UT: Gibbs Smith, 2000.

Sandra Markle, *Outside and Inside Rats and Mice*. New York: Simon and Schuster, 2001.

Bonnie Highsmith Taylor, *Lucy: A Virginia Opossum*. Des Moines, IA: Perfection Learning, 2001.

Patricia Whitehouse, *Rats*. Chicago: Heinemann, 2002.

Amy Bartlett Wright, Jonathan Latimer, and Karen Stray Nolting, *Caterpillars*. New York: Houghton Mifflin, 2000.

Glossary

anxiety: extreme nervousness

camouflage: physical features that let an animal blend into its environment

catatonic: lack of movement or expression

marsupials: animals that carry their young in a pouch, such as kangaroos and oppossums

molecule: the smallest particle of a substance

pheromones: chemicals produced by animals that stimulate or attract other animals

primate: the family of mammals that includes humans, apes, and monkeys

radar: a device that locates objects by sending out radio waves that the objects reflect

rodents: small mammals such as rats and mice

sonar: a device that locates underwater objects by sending out sound waves that the objects reflect

Index